The Other Presidency

Th: Jefferson

The Other Presidency:
Thomas Jefferson and the
American Philosophical Society

Patrick Spero

With a foreword by Andrew O'Shaughnessy
and an afterword by the author

The American Philosophical Society Press
Philadelphia

Published by
The American Philosophical Society Press
Philadelphia, Pennsylvania 19106-3387
www.amphilsoc.org

Printed in the United States of America on acid-free paper
10 9 8 7 6 5 4 3 2 1

Library of Congress Control Number: 2024931562

Paperback ISBN: 978-1-60618-904-7

Hardcover ISBN: 978-1-60618-903-0

eBook ISBN: 978-1-60618-905-4

Contents

v

Foreword

Andrew O'Shaughnessy

The breadth of Jefferson's intellectual interests and his desire to disseminate knowledge were apparent in his forty-six-year involvement with the American Philosophical Society (APS). Founded in 1743 by Benjamin Franklin and others, its mission was to promote useful knowledge. Like the Royal Society in London, such scientific societies often acted as the principal conduits for research and inquiry, rather than universities and colleges, whose primary focus at the time was teaching.

In *The Other Presidency*, Patrick Spero aims to relate and assess the role of Thomas Jefferson at the American Philosophical Society. From the time Jefferson became a member in 1780, Spero makes the case that Jefferson's impact upon the Society was highly significant. Especially during his period as vice president of the United States, Jefferson set the agendas at APS, presented his own research papers, encouraged other members, donated books and artifacts, and extended the Society's network, including admitting international scholars. While Jefferson was vice president, the

Society held an essay competition "for the best system of liberal education and literary instruction ... comprehending also a plan for instituting and conducting public schools in this country." The book demonstrates the eclectic interests and curiosity of Jefferson and offers an excellent window into his intellectual life.

In 1797, Jefferson became president of the American Philosophical Society, which he described as "the most flattering incident of my life." As its president until 1814, Jefferson still holds the record for his tenure in this office. While the Society's president, he presented his own scholarly papers, including one in 1797 on the bones of a large animal recently excavated in western Virginia, that he called a *Megalonyx*, and another paper, the following year, in which he discussed his own design for a moldboard plough. He influenced the research and collecting agenda of the Society by encouraging the library to collect information about the past and present state of the country, showing a particular interest in artifacts and information relating to Native Americans. As secretary of state and later vice president of the United States, Jefferson continued to attend the Society's meetings, usually held at least once a month, and often lasting several hours. He donated books, manuscripts, and artifacts to the Society, including the journals of Lewis and Clark, having encouraged Meriwether Lewis to consult with the Society in 1803 in preparation for their transcontinental journey. Owing to his preference for independent civil organizations that were free of central government control, he commissioned the Society to undertake initiatives of national importance which included support for the Michaux Expedition, the collection of artifacts and specimens of natural and Native history of America, and involving the Society in advising the Lewis and Clark Expedition.

Spero demonstrates that during the 1790s, the membership of the Society was increasingly composed of Jeffersonian Republicans. Furthermore, he shows that this was particularly true of the most active members. He suggests that this might reflect the natural inclinations of the scientific community, rather than deliberate political partisanship, although the latter should not be discounted since Jefferson was clear that he aimed to weaken the influence of the Federalists when he founded West Point and the University of Virginia. In any case, the book finds that the leadership positions in the Society had become more bipartisan by 1810.

Although this book is intended primarily as an overview that details Jefferson's involvement with the APS, Spero offers new insights and arguments, the most important of which is the discussion of the political influence of Jefferson on the APS. As opposed to the manufacturing emphasis of the Federalists, Jefferson preferred research in natural history, in part because he believed that the findings would support and implement his agrarian vision of America.

Thomas Jefferson, painted by Thomas Sully in 1821.
Courtesy of the American Philosophical Society.

The Other Presidency:
Thomas Jefferson and the
American Philosophical Society

L et us begin with the simple facts. In 1780, the American Philosophical Society elected Thomas Jefferson to its membership, the beginning of a relationship that would last until Jefferson's death in 1826. During those 46 years, Jefferson served as a member of the Society's Council (its governing board), held the office of vice president from 1793 to 1795, and finally was its president from 1797 to 1814. His election to the APS presidency, Jefferson remarked, was "the most flattering incident of my life," and he held onto this appointment even while serving as vice president and president of the United States. After resigning from the APS presidency in 1814, he continued to stay involved in Society business through an extensive correspondence network,

This essay was previously published by the American Philosophical Society for distribution at the November 2018 Annual Meeting on the occasion of the Society's 275th anniversary.

as an elected Councilor from 1818 until his death, and by contributing important collections, nominating new members, and providing general guidance to Society officers and committees that ran the Society's affairs. Needless to say, on at least this superficial level, the APS was a large part of Jefferson's life.[1]

Biographers, however, have largely overlooked this aspect of Jefferson's life. Merrill D. Peterson's magisterial biography—with over 1,000 pages of text—mentions the APS only a handful of times, and offers no sustained analysis of its role in Jefferson's life.[2] Joseph J. Ellis only mentions the APS once in his award-winning *American Sphinx* when he notes Jefferson's election to membership.[3] Jon Meacham's much longer study does only slightly better with two mentions, one on Jefferson's election and another on the institution's role in supporting a scientific expedition that Jefferson organized in 1793.[4] John B. Boles, Jefferson's most recent biographer, follows this general pattern, mentioning the APS seven times, mostly to point out that it was an institution to which Jefferson belonged and to which he occasionally contributed items of scientific interest.[5]

More specialized studies of Jefferson as scientist and intellectual do slightly better than the general biographies, but not by much. Kevin J. Hayes's *The Road to Monticello: The Life and Mind of Thomas Jefferson* has more than a dozen references to the APS, but only to mention it as an organization with which Jefferson was involved.[6] Silvio A. Bedini's exhaustive *Thomas Jefferson: Statesman of Science* has the largest number of APS mentions, but these too are mostly passing references.[7] Perhaps not surprisingly, the former Executive Officer of the APS, Keith Thomson, offers perhaps the most in-depth descriptions of Jefferson's relationship to the APS in his book, *Jefferson's Shadow: The Story of His Science*. But even his account serves more as a summary of Jefferson's most well-known scientific contributions to the Society, such as his essay published by the Society on the fossil known as the *Megalonyx* or Jefferson's donation of Lewis and Clark materials to the Library's collections, than as a detailed account of Jefferson's relationship to the APS.[8]

Although it is impossible to explain conclusively this consistent oversight, there are at least a few possible reasons. First, it is entirely possible that Jefferson's involvement with the APS was purely superficial, that he may have held several offices but they were mere sinecures. Jefferson's comment that his election to the presidency was "the most flattering incident of his life" may in fact have been the type of edifying flattery that politicians offer upon receiving any honorary designation. Biographers may have determined—consciously or subconsciously—that the APS's role in Jefferson's life, and his role in the institution, was a minor incident in a life of greater significance. Such a relatively unimportant engagement would indeed only merit the passing references to Jefferson's involvement with the APS that his biographers universally offer. Another possible explanation, perhaps the most likely one, is simpler. An analysis of Jefferson and the APS is not a self-evident line of inquiry for a biographer to follow. In other words, biographers have ignored Jefferson's relationship with the APS because no one thought to ask the question.

The APS, for its part, has taken a different tack. The Society has proudly touted its connection to Jefferson. Jefferson is, put simply, everywhere at the APS. There is an elegant garden next to Library Hall named after him. One of the Society's most prestigious awards carries Jefferson's name. The APS Museum recently mounted three major exhibitions in his honor. A Sully portrait adorns the main stage on which all major APS meetings are held, and the Houdon bust rests safely in the Society's vault.

Amid all of this adulation, however, the institution has in another way followed the route of Jefferson's biographers and done relatively little to uncover the depth of Jefferson's relationship to it. As far as we have found, there have been only two substantive engagements with the topic by the Society. In 1943, on the 200th anniversary of Jefferson's birth and the APS's founding, APS member Gilbert Chinard published "Jefferson and the American Philosophical Society" in the Society's *Proceedings*. Chinard's essay is a wonderful documentation of the offices Jefferson held and his main correspondence with the institution. As Chinard admitted, however, his "paper purports only to be a rapid sketch and a brief survey of a subject deserving a more complete documentation and a more elaborate presentation."[9]

The next major contribution came in 1997, the 200th anniversary of Jefferson's election to the APS presidency, when APS Librarian Edward C. Carter delivered a speech, subsequently published as an APS pamphlet, at a conference co-sponsored by the APS and Monticello on Jefferson, education, and religion. Carter added to Chinard's work by uncovering new information on Jefferson's early history with the APS. Carter found that Jefferson had attempted to contribute a meteorological diary to the Society's holdings as early as 1779, a year before he was elected a member, and also attended virtually every APS Meeting in Philadelphia while he was secretary of state and vice president in the 1790s. Yet, although both scholars helped establish the extent of Jefferson's official involvement in the APS, neither offered an examination that went much beyond institutional history.[10]

No study, as far as we have found, has examined the significance of the APS in Jefferson's intellectual life, nor has anyone studied Jefferson's influence on, even leadership of, the institution's development in the Early Republic, which took it from a fledgling Philadelphia outfit into the nation's leading scientific organization. Thanks to newly digitized databases that provide even greater access to both Jefferson's correspondence and the APS's own institutional records, however, a closer examination of Jefferson's involvement in the Society sheds new light on both

Jefferson's intellectual interests and the APS's development during Jefferson's lifetime.[11]

The evidence suggests that Jefferson's association with the APS went through three distinct phases: the Philadelphia period (1791–1800), the American presidency (1801–1808), and his retirement (1809–1826). In each period, Jefferson's own place in life and his needs of that moment determined his interactions with the Society. Conversely, the Society sought in Jefferson different things as his career and life evolved. Throughout this mutually constitutive but always shifting relationship, a few imperatives guided both Jefferson and the APS. First, Jefferson and the APS generally focused on the natural history of America—that is, a study of the continent, its environment, geography, and flora and fauna. They both also shared a desire to use the work of the APS to establish the reputation of American science on par with their European counterparts. Finally, they saw their scientific endeavors as advancing the nation's interests, often in ways that were meant to aid Jefferson's own view of making the United States an expanding republic built upon independent yeoman farmers. This book is an attempt to document and analyze this important relationship.

When Jefferson first learned of the APS remains something of an open question, although its operation was surely known to Jefferson years before his first correspondence with the Society in 1779. The Society's founding dates, rather serendipitously, to Jefferson's birth year. Indeed, it occurred a month after his April 13 birth. On May 14, 1743, Benjamin Franklin and several other public-spirited men began to organize the American Philosophical Society by publishing *A Proposal for Promoting Useful Knowledge.* Modeled on the Royal Society, their "American Philosophical Society" was to advance knowledge through a class of members elected based on their demonstrated learning. These members were to share new knowledge with each other through regular Meetings and through a correspondence network that would stretch across the Atlantic.

The Society was also to support new research through direct funding of experiments.[12]

Although an inspired idea, the institution languished in the early years. It had become so lethargic by the 1760s that a group of young upstarts created a second organization in 1766 called "The American Society Held at Philadelphia for Promoting and Propagating Useful Knowledge," which took its motivation (and part of its title) from Franklin's original proposal. This new group made clear in its mission statement that it was formed with a particular and very practical purpose: "the advancement of useful knowledge and improvement of our country."[13]

In 1768, in a move that would come to define the institution that Jefferson would later embrace, the more venerable if quiet American Philosophical Society and the more recent if energetic American Society Held at Philadelphia for Promoting and Propagating Useful Knowledge merged to create "The American Philosophical Society, held at Philadelphia, for Promoting Useful Knowledge," a long title that remains its official name today. The political crisis then riling the colonies appeared to have influenced the partnership. As colonists headed toward revolution, they joined together in boycott movements and created a raft of associations meant to offer a unified response to imperial policies. Amid this moment of unity, the comparable missions of the two bodies seemed to undermine this effort at Colonial cooperation, and the institutional rivalry between members of both organizations seemed to dissipate as the political crisis escalated.[14]

The new body's purpose-driven mission embraced that of the newer Society, although it gained credibility by accepting into its membership the more august members of the American Philosophical Society, such as Benjamin Franklin. The APS would, as noted in its rules, support science that served the public, writing:

When any useful discoveries are made, either by new Inventions or by the improvement of the old, these shall be published by the Society in the plainest and most intelligible manner, and pains taken to introduce them into common practice, that all may reap Benefit from them.[15]

The newly constituted organization, as embodied by the above rule, was shaped by and tried to influence the historic moment in which it was born. With many colonists feeling slighted by Great Britain, the Society wanted to showcase the ability of colonists to operate on par with their British colleagues. David Rittenhouse's observation of the transit of Venus in 1769, a project supported by the APS, was a clear example of the Society and its members attempting to establish Americans on a level equal to—if not superior to—European scientists. Rittenhouse's careful observation of the transit created an international sensation. His recordings and those of others in America served as the basis for the first edition of the APS's *Transactions,* published in 1771. Rittenhouse's findings became required reading for any scientist and established the APS as a legitimate scientific institution among its European peers of longer standing. As one of Jefferson's friends noted to him, Rittenhouse's success "procured their Society so much Honour abroad."[16]

In a related vein, the Society used its resources to encourage experiments meant to address practical concerns that colonists had about their status in the British Empire. One of the chief complaints then emanating from Colonial cities was that strict imperial regulations had limited Colonial manufacturing and slowed their economic growth. Society members undertook an initiative to resolve this problem by using their financial and intellectual capital to incubate a nascent silk industry in Pennsylvania. Adding a new commodity to the colonies' panoply of staple crops, the thinking went, would improve the Colonial economy while also serving the Empire's goal of keeping the colonies producers of raw materials and the home country the manufacturing hub that turned these items into consumer goods for re-export.[17]

To launch this agricultural experiment, the Society's aptly named "Committee for American Improvements" created "the Silk Society," a private enterprise of which the Society was the chief shareholder. The Society eventually raised a significant sum of money to import silk cultures and hire managers for the initiative. They also published *Directions for the Breeding and Management of Silkworms*, a guide that they hoped would encourage individuals in other colonies to develop their own industry. As their *Directions* made clear, this APS-backed foray was no mere business enterprise. It was as much a scientific experiment as it was an entrepreneurial undertaking. The publication's goal also served the Society's new mission to encourage the dissemination of useful knowledge throughout American society, noting in their publication that regular colonists could easily incorporate silk production into their homes if they followed the Society's instructions. Although the project proved to be something of a failure, the imperative behind it, that is, the belief that the APS should support scientific endeavors that promised to address real-world issues and improve American society, would continue to guide the Society's activities into the Early Republic.[18]

Jefferson and others in Virginia were clearly aware of and inspired by the APS and its undertakings in this period. Indeed, in the 1770s, probably around 1773, on the heels of the *Transactions* volume and the creation of the Silk Society, Jefferson and some of his Virginia compatriots formed the Virginia Philosophical Society for the Advancement of Useful Knowledge, an organization modeled on the APS. Their purpose was similar to their Philadelphia counterpart, declaring the practical reasons for their scientific research in a newspaper article:

> Who can tell what may accrue to the Inhabitants from an Acquaintance with the Nature and Effects of the Climates and Soils? The minerals, fossils, and Springs, in which the Country abounds, may yield the greatest Emolument both to their owners and the Publick. The Multiplicity of Vegetables and Animals may conduce to the Purposes of Commerce and the Comforts of Life, in Modes with which, at present, we are not acquainted.[19]

While the Virginia institution never got much off the ground (in fact an unfortunate fire destroyed their building and collection of artifacts), it did support endeavors similar to the APS's Silk Society. Moreover, its creation shows that Jefferson saw the need for institutions to bring together like-minded scientists to study the natural world and promote knowledge in order to serve the public. Indeed, this use of science as a means to advance the interests of America became a key element of Jefferson's vision in the years to come.[20]

Although Jefferson was elected to the APS in 1780, part of a bumper crop of members that also included George Washington, Alexander Hamilton, and John Adams, and served as a Councilor from 1781 to 1782 and 1783 to 1785, his real involvement with the APS began in 1790 when he moved to Philadelphia as secretary of state and, in 1791, became vice president of the APS. The Society itself was undergoing a transition when Jefferson arrived in the temporary capital in 1790. After a period of greater activity during the imperial crisis, the divisions of war tore apart the Society and its members. Meetings adjourned for much of the 1770s, but with military victory in the offing, the Society began to reawaken just as the nation was finding its own footing. Members started to meet again and rebuild the Society's roll, as shown with the election of 1780. By the time

Jefferson arrived in 1790, the APS was carving out a reputation as the leading scientific organization in the new nation, one that continued the tradition of the earlier Society to serve the public good through the pursuit of knowledge.[21]

Jefferson found in the newly energized APS a safe haven during his otherwise tumultuous stay in the City of Brotherly Love (1790–1794, 1797–1800). According to the APS Minutes, Jefferson attended 32 Meetings, making him one of the Society's most devoted attendees. As Society Librarian Carter noted, in 1798, "perhaps the most difficult year in Jefferson's life," Jefferson was at every Meeting held while he was in Philadelphia. Meetings occurred at least once a month, and more often twice a month, and lasted several hours. Although they inevitably involved food and drink, the central feature of the gatherings were prepared talks about the latest research that members had either conducted or learned of through correspondence or books. From 1797 to 1799, after being elected president of the Society, Jefferson often presided over the Meetings he attended. Indeed, even though he was vice president of the United States at the time, it is likely that he dedicated more of his attention to his duties as APS president than to his duties as vice president.[22]

But it was more than simply a convenient social club. The research of the members and the experiments the Society funded often fed Jefferson's own interests in natural history. Many of the Society's most active members, such as Benjamin Smith Barton, William Bartram, and David Rittenhouse, were prominent natural historians and scientists whom Jefferson admired. Indeed, influenced by the work of naturalists like these, Jefferson had already undertaken research that mirrored the Society's own focus on exploring the natural world. During the 1770s, he kept a detailed meteorological diary at Monticello, and, in the 1780s, he began conducting experiments on plants in his gardens, sometimes exchanging seeds with John Bartram. Finally, in perhaps his greatest scientific undertaking, Jefferson began compiling Native American languages in the 1780s in hopes that a linguistic analysis of them would reveal new information on the historic development of North America and its peoples, a subject that the Society would come to embrace under Jefferson's leadership.[23]

If Jefferson found an intellectual home in the APS that nourished his interest in natural history, the Society found in Jefferson a figure who could help it better establish its overall direction and reputation in the Early Republic. On this front, Jefferson, in his leadership positions as vice president and president, played an active part in setting the agenda for Meetings, often using his correspondence network to shape an evening's program, which in turn established the tenor of the Society's research program more generally. In total, Jefferson made at least twelve presentations at Meetings. His contributions reflected his eclectic interests. For instance, at a Meeting in August 1791, Jefferson presented a paper he received from a Mr. John Cooke of Tipperary in Ireland about a new standard of weights and measures, a subject that Jefferson wrote about extensively in the 1790s. At the same Meeting, Jefferson also presented a "curious piece of Indian Sculpture supposed to represent an Indian Woman in Labour" that had been found in Kentucky and sent to Jefferson by Henry Innes, a federal judge in that state. In 1798, Jefferson read a paper on a new model for a threshing machine that he received from Thomas Martin only a month before. And the following month, May 1798, Jefferson read a paper detailing his own new design for a mouldboard plough. Although these are only a few examples of the many books, papers, pamphlets, specimens, and artifacts that Jefferson presented in person or sent via friend or post to the Society, they help show one of the ways Jefferson's leadership shaped the Society's priorities in the Early Republic.[24]

The scientific content was surely a draw for Jefferson, as was the conviviality of the Meetings themselves, but there was more to Jefferson's interest in the APS than science for its own sake. There were always subtle but significant political elements to Jefferson's scientific activities, and these political purposes also comported well with the APS's culture in the 1790s. Jefferson's involvement with the APS in the 1790s only increased as the partisan atmosphere in Philadelphia grew hotter, and these two occurrences are no simple coincidence. With Federalists arguing for an energetic federal government to lift the young nation, Jeffersonians feared that this governing model was simply a return to the state-centric, hierarchical European form of governing that the democratic American Revolution was meant to displace. For Jeffersonians, the Federalists' desires for centralization threatened the vitality of independent organizations, like the APS, which tried to undertake initiatives that might be in competition with the government.[25]

The most notable example of this contrast was Hamilton's attempt to use the federal government to directly aid industries through bounties and by creating a "Society for Establishing Useful Manufactures," something akin to the APS's Silk Society. In this case, however, the government established the organization and provided it with significant benefits. Although Jefferson supported the idea of more efficient manufacturing, especially innovations that supported farmers, Jefferson saw Hamilton's use of the federal government to provide direct aid to manufacturers as an inappropriate use of government power. In Jefferson's view of government and society, local bodies, not the federal one, and civic institutions, like the APS that were controlled and supported by American citizens devoted to improving the nation but dependent only on themselves, were the proper means of supporting innovation, experimentation, and industry. As Jefferson wrote in notes he penned in opposition to Hamilton's famous *Report on Manufactures,* he believed such public support for industry was unconstitutional. But, more important, he thought it unwise. State governments, he believed, were best positioned to determine the industries likely to thrive in their local environments. But even in allowing for some state support, Jefferson remained cautious. There were "few instances" of such publicly funded outfits "being successful," he found. Indeed, he concluded that they often could only sustain themselves by "burthening the public."[26]

The APS and its mode of operating in the interests of the nation but independent of federal aid thus offered Jefferson a powerful alternative to Federalist ideas. Indeed, in 1791, Jefferson himself undertook an expedition under the auspices of the APS that served this exact purpose. A pest called the *Hessian fly,* so-named because Americans suspected that German soldiers brought the insect to American shores during the Revolution, was wreaking havoc on American crops, especially the wheat crops of New York and parts of New England.[27]

The threat of an agricultural catastrophe spurred the APS to act. Combating the spread of the Hessian fly struck at the core of the APS's mission to use science to solve a problem that threatened the vitality of the nation. Jefferson was especially interested in preventing the fly from reaching Virginia, where wheat was fast replacing tobacco as the crop of choice. The Society created a committee with Jefferson at its head to study the fly and its history, all with an eye toward halting its spread. They sent out circulars and requested information from those in their correspondence networks.[28]

In 1791, in service to the APS, Jefferson, along with his friend and fellow APS member James Madison, headed to New England to further study the natural history of the infestation.

His journey took him to New England to chart the course of the infestation, leading him to conclude that it began around 1776 in New York, adding further evidence that it was brought over during the Revolution, potentially by Hessian soldiers. Critics of Jefferson at the time claimed his scientific expedition was a front for partisan purposes, but Jefferson's research remains, in the words of one historian, "the most detailed account ever written of the Hessian fly's early migration patterns."[29]

Related to the question of whether government should support science was the equally charged question of how America would achieve its independence from Europe. Here, too, the parties were divided on the issue, and scientific research once again became a flashpoint. Federalists, of course, wanted America to be an independent nation, meaning that the country stood on at least equal footing with European nations, ultimately reliant on its own economy and people for its growth. They also felt, however, that it could only do so by first engaging in trade and exchange with Europe, especially Great Britain. Implicitly, the Federalists recognized that the United States' position in the Atlantic World was still that of an agricultural nation dependent on trade, financing, and innovation from Europe. Their focus for research was thus based on improving manufacturing technologies, often those imported by Europe, as exemplified by Hamilton's Society for Establishing Useful Manufactures and Washington's own importation of an Oliver Evans mill in 1791. Federalists, many of whom were based in seaports, especially in New England, also seemed to emphasize mathematics and astronomy in their schools, both fields that aided in oceanic travel, rather than the natural history that interested Jefferson and the APS.[30]

Jefferson abhorred the Federalist view of American independence through trade and industrialization, and instead wanted science, especially research in natural history, to reveal the American environment's potential, its grandeur, and its distinctiveness, all of which would support his vision of an expanding, agrarian republic. Americans, he believed, could achieve their own independence on their own terms by harnessing the power of the plough, rather than following the model of industrial development that Europe, and especially Great Britain, had adopted. To do so, however, Americans needed to master this environment, fueling Jefferson's interest in natural history to serve his nationalistic vision of a republic founded upon agriculture. Scientific discoveries in the natural world, he believed, could encourage the success of small farmers whose dependence on their own hands to till the soil would allow them greater personal freedom than industrialized wage laborers.

Indeed, one of Jefferson's own inventions in this period, one presented to the APS, was an improved mouldboard plough that he hoped would increase the efficiency of farms and improve the nation's agricultural production. To Jefferson, then, natural history was essential to the future of an expanding country of farmers. While Jefferson had interest in astronomy and mathematics, the APS, with its inclination toward natural history, was the natural outlet for Jefferson to pursue that field of science in support of his vision for an agrarian republic that relied on the control of the environment for its success.[31]

In addition to the practical purposes of greater efficiency, Jefferson thought that more scientific research on the natural world would aid in building the new government and should inform its function. America's climate, he pointed out, required new types of building materials that were able to withstand the continent's weather, and the land itself provided different raw materials from which to design structures. These environmental differences with Europe, Jefferson felt, also had implications for governing. As historian Linda K. Kerber so aptly summarized, Jefferson believed deeply that "the American land could sustain a viable state and, in fact, a new civilization." Indeed, he went so far as to speculate that the country needed to construct its own set of governing principles suited for the American setting, writing to an APS member in 1800 "that our geographical peculiarities may call for a different code of natural law to govern our relations with other nations from that which the conditions of Europe have given rise to there." Thus, underlying Jefferson's scientific emphasis on the natural world—as opposed to the Federalists' astronomical and mechanical focus—was a political project premised on creating a distinct American state shaped by a greater understanding of its environment.[32]

The American Philosophical Society became Jefferson's means for advancing this vision through science. Two of Jefferson's major scientific undertakings in this period fit this agenda, neither of which would have been possible without the APS. The first was the ill-fated Michaux Expedition. The story of this undertaking begins in 1793, when a French envoy to the United States named Edmond Genet arrived on America's shores with the goal of gaining American support for the French cause against the British. He came to the United States just as the First Party System was forming, and soon he would become a central part of it. In the 1790s, Federalists and Jeffersonians were dividing over the meaning of the French Revolution, with many Jeffersonians seeing it as an extension of the American Revolution, and Federalists a deviation from it. Because of that, Federalists suspected that Genet was a spy meant to undermine their government so the Jeffersonian Party could take over.[33]

Genet's diplomatic mission became a partisan issue once Genet started to interfere in a whole range of American affairs, which only confirmed Federalists' fears of his intent. Among the things Genet did was to become a champion of the scientific ambitions of French botanist André Michaux. Michaux had been traveling throughout America since 1785 on a royal commission from the French Crown to collect and observe the flora and fauna of the Eastern United States. Hearing that the APS had previously contemplated a plan to explore the Western interior, Michaux proposed that the Society sponsor him to undertake the journey. Jefferson found the proposal inspiring. It reflected Jefferson's long interest in exploring the West for both scientific purposes and as reconnaissance for America's eventual conquest of the territory, something that was necessary for Jefferson's vision for an expansive and agriculturally based country composed of independent yeoman farmers.[34]

Jefferson took the lead in organizing the endeavor and, although it was a mission of national import, he knew that the American Philosophical Society was the institution through which it should be run. As one historian remarked, Jefferson knew that "it was a matter for the American Philosophical Society, not the American Government." Jefferson drew up detailed instructions for Michaux, including the recommendation that he tattoo his most important findings on his skin so they would not be lost. He also drafted a proposal for APS members to raise the funds necessary for such an ambitious undertaking. Jefferson raised $870 in his campaign, and the APS continued the effort, raising an additional $699 for a total of $1,569 for Michaux's expedition (for comparison, the US Congress allotted Lewis and Clark $2,500 in 1803). The proposal received wide, bipartisan support. Prominent Federalists Washington; Hamilton; vice president of the United States, John Adams; and secretary of war, Henry Knox were among the donors, for instance.[35]

Michaux's ambitious voyage, however, ultimately faltered because of the politics surrounding Genet. Although Jefferson initially viewed the expedition as a purely scientific endeavor, Genet gave Michaux additional directions meant to serve the French interests, instructions that turned his mission into a quasi-diplomatic expedition for France. After only a few months, the

Washington administration requested Genet's recall because of the controversy that followed him. Michaux's expedition, meanwhile, became embroiled in Genet's schemes and faltered. But even if the expedition failed, Jefferson's intent, and the APS's strong backing of the project, reflected the priority they gave to exploring the interior areas of America so they could be understood and eventually mastered.[36]

Jefferson's involvement in the APS's "Committee to Collect Information Respecting the Past and Present State of This Country" constituted his second major scientific initiative at the APS, which was meant to support a nation-building project on Jeffersonian terms. The committee was formed in May of 1797 at the first Meeting Jefferson presided over as APS president, and Jefferson was officially appointed to the committee a year later in 1798. Jefferson and the other members of the committee met for almost two years to discuss how the APS should develop a collection of "American antiquities." They ultimately determined that the APS should become a repository for America's natural history and Native past.[37]

When the committee's deliberations concluded in 1799, the Society sent out a circular letter under Jefferson's signature as president to announce this goal and encourage people to send information to the APS. The letter opened with a clear statement of the APS's national purpose and interests. The Society, they wrote, "have always considered antiquity, changes, and the present state of their own country as primary objects of their research." The collecting goals they outlined reflected their desire to learn of the natural and human history of the continent as well as learn more about its current geography and climate. They aimed "to procure one or more entire skeletons of the Mammoth," "to obtain accurate ... descriptions ... of ancient fortifications, tumuli, and other Indian works of art," to receive descriptions of rivers, mountains, and lakes, and to acquire ethnographic and linguistic reports about Native peoples. The Society sent the circular to "gentleman who have taste and opportunity for such researches" throughout the country. The committee soon received various reports from the field, most dealing with geographic observations and information on Native American cultures, and they started to compile this correspondence in the Society's Library, adding manuscripts to its collection of books.[38]

Jefferson's goal to make the APS the collecting center for information on the continent's history was guided by his vision of American exceptionalism. Just as Jefferson wanted to study the natural world to understand the distinctiveness of America through expeditions like Michaux's, so too did he hope that by acquiring a collection of ancient specimens and Native texts in one place, the Society would uncover the rich and deep past of the continent that would prove the North American continent had its own unique history that was relevant to the present. This past, too, was meant to prove America's separateness from Europe. The findings, he hoped, would inform the way the nation understood itself and its history.[39]

His interests in proving the exceptionalism of the American continent were fueled, of course, by his famous and protracted transatlantic debate with the French scientist and international APS member Georges-Louis Leclerc, Comte de Buffon, about the virility of the American environment, a debate that also shaped the APS's scientific and collecting agenda. Buffon had posited that the American climate was degenerative, that it was wetter and colder than Europe and therefore caused species to be smaller, weaker, and inferior, a hypothesis that extended to its Indigenous peoples as well as the burgeoning American nation. Jefferson took umbrage with this argument. Jefferson not only believed that the science was bad, but the implications of the argument insulted his sense of national pride, as Buffon's logic implied that the United States would remain inferior to European nations. Jefferson thus looked to science to show that Buffon was right in assuming that America was distinct from Europe. But unlike Buffon, he intended to show that its environment was one of strength, and even superiority to Europe, thus his desire for the APS to recover a full mammoth skeleton to prove that the American continent was home to massive creatures.[40]

In fact, Jefferson's interest in Native American culture was in part influenced by Buffon's claim that Native Americans were inferior to Europeans. He made a point to defend the strength and intelligence of Indians in his *Notes on the State of Virginia*, a work he first considered publishing through the APS. When it came to his section on Native American cultures, he stated that their bravery in war was unmatched, and compared their oratory skills to those of the great classical orators Cicero and Demosthenes. This evidence was important for Jefferson to show that North American peoples whose roots predated European contact had as much capacity as those in Europe, and, thus, the new American nation and its citizens who inhabited this same environment had the potential to replicate—even surpass—the classical models that Europeans often held up as ideals. If he conceded anything to Buffon, it was that if Native Americans displayed any physical inferiority to Europeans, it was a result of their diet rather than any inherent deficiency.[41]

After the publication of *Notes*, Jefferson continued to investigate Native American cultures with the expectation that his findings would provide further evidence that Native Americans were a complex people with a distinguished history and culture. Jefferson's line of inquiry aimed to buttress the notion of Native Americans as "noble savages," a contrast to Buffon's claim that "the savage is feeble." Of course, another pressing reason for Jefferson's desire to document Native cultures was that his vision of a westward-expanding nation meant that Native peoples and cultures would likely be displaced, perhaps even erased, a conceit he noted to his friend Benjamin Smith Barton in 1809. After losing most of his own Native fieldwork to theft, a depressed Jefferson confided to his friend that "my opportunities were probably better than will ever occur again to any person having the same desire" in part because of the destruction of Native cultures occurring in the face of colonization. He thus saw the APS as the institution through which he could preserve a part of America's Native past.[42]

The desire to excavate—often literally—America's natural history was imbedded in the APS's agenda from its outset, so its inclusion in its collecting agenda is no surprise. But Jefferson made the field an even more pressing concern of the Society in 1797 when he gave a paper at an APS Meeting on the bones of a large animal dubbed the *Megalonyx* that were recently unearthed in western Virginia. On March 10, 1797, Jefferson presented a detailed report on the bones. Jefferson's initial assessment of the Megalonyx was that it was a massive, carnivorous tiger, something Jefferson was excited to share with his intellectual adversaries who subscribed to the ideas of Buffon. Further study by his friend and fellow APS member Caspar Wistar, however, led him to conclude that it was a giant, plant-eating sloth. Even so, Jefferson used the discovery to argue that these bones proved that a giant creature inhabited the North American continent. At the time, Jefferson speculated that the animal might still roam in the West, and he hoped that this find would disprove Buffon's thesis by showing that large, fierce animals inhabited the American terrain. Jefferson eventually used the *Transactions* as the vehicle to disseminate his theory, further establishing the APS's publication as the preeminent scientific journal of the country. His findings caused a sensation when published and remain an event of note in the history of science as the founding moment for the study of vertebrate paleontology in America. Inspired by Jefferson's work, the Society continued to support the search for big bones by making it a key part of the institution's official collecting strategy in 1799.[43]

Jefferson and the APS thus had a symbiotic relationship while Jefferson resided in Philadelphia. Jefferson found in the APS an institution through which he could conduct research, and the APS found in Jefferson the leader who would help establish the institution's reputation in early America. By the time Jefferson left Philadelphia in 1800, he had given new direction to the Society. He served the national interest through the Society's work on the Hessian fly, and he continued to do so in other ways, especially through his work at the APS on establishing standards for weights and measures and through the support of a local viniculture industry, a project similar to the earlier Silk Society. As exemplified by funds he raised for the Michaux Expedition, he had made the APS a supporter of ambitious scientists and their projects. Through the committee on collecting, he established the APS as a leading repository and carved out some of the core areas of collecting that still guide its collection development today. Finally, through his contacts at home and around the world, he shared cutting-edge research at the APS and used his network to strengthen the APS membership by recommending those who impressed him the most.[44]

Jefferson's stay in Philadelphia coincided with the birth of the First Party System, one of the most divisive political moments in the country's history. With Jefferson as the symbol of both the opposition party and the American Philosophical Society, the APS became, perhaps unwittingly, a part of the partisan environment that consumed the city. Indeed, during this period, the institution took on a decidedly Jeffersonian cast, something that Jefferson may have encouraged but most likely was a natural outgrowth of the inherent political sympathies of Philadelphia's scientific community at the time. In several cases, the APS elected members who began as friendly correspondents of Jefferson and served his interest in natural and Native history. For instance, Jefferson corresponded extensively with William Dunbar, Robert Livingston, John Stuart, and Louis Hue Girardin, all before they were elected to the APS. In the case of Stuart, Jefferson used the promise of APS membership as an enticement for Stuart to send information about the Megalonyx to him, telling Stuart that such information would make him "a person so worthy as ... to be taken in to their body."[45]

The election of Jefferson's contacts to the APS thus only added reinforcements to what appears to have been an already sympathetic membership. Although the Society elected members independent of party, a closer examination of the APS Minutes shows that the most active members—those who attended Meetings and determined policy—were skewed toward the Jeffersonian Party. In a sampling of 35 Meetings in the 1790s, not all of which Jefferson attended in person, members with Republican leanings outnumbered Federalists by approximately 3:1 and outnumbered those with no discernable political affiliation by a factor of 3:2. Aside from local residents, the international membership of the Society seemed to be colored by partisanship as well. Where Federalists wanted to strengthen connections with Great Britain and undertook initiatives to establish clear ties across the Atlantic with Britons, Jefferson and his supporters saw the French as America's closest ally and shunned British ties. Likewise, between 1775 and 1800, the Society elected 55 French members and only 33 British members, even though this period coincided with an enormous efflorescence of discovery coming from Great Britain's scientific community.

The Jeffersonian cast of the Society becomes even more evident in the partisan makeup of its officers. Over the course of the period from 1791 to 1800, the height of Jefferson's direct and personal engagement with the Society in Philadelphia, as well as the birth of the First Party System in which Jefferson served as the figurehead of the opposition party, the Society elected over twice as many Republican-affiliated officers as Federalist ones (Figure 1). Jefferson's predecessor as APS president, David Ritten-

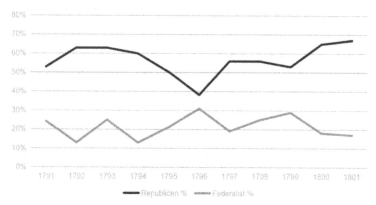

Figure 1. Partisan composition of the elected officers of the APS, 1791–1801.

house, who was elected to the office after Franklin's death, had distinct Republican sympathies, as did other long-serving officers like Caspar Wistar, Robert Patterson, Benjamin Smith Barton, Charles Willson Peale, and Benjamin Rush.

This combination of science and politics embroiled the Society in political controversy. Federalists in the 1790s soon saw Jefferson's scientific interests in natural history and the institution that embodied them as political vulnerabilities that were ripe for attack. Where Jefferson saw his science as a part of his nation-building project, Federalists considered it a waste of time, one that was—somewhat ironic considering Jefferson's disdain of Europe—more suited for the theoretical realm of European intellectuals than the needs of the country. Federalists saw natural history, as Linda K. Kerber noted in her study on the politics of science, as "a gentlemanly and amateurish curiosity about the plants that grew on one's farm." The hunt for mammoth bones became a particularly enticing target of ridicule. It was, Jefferson's opponents claimed, a pointless pursuit that had little relevance to the present. Jefferson's close affiliation to the APS, and the institution's support of natural history during this partisan moment, made it an easy target of Federalist barbs. "The crude syllogism," Kerber concluded, "seemed to run: Democrats are suspect. The American Philosophical Society is largely Democratic, and concentrates on natural history. Therefore there is something suspicious about natural history."[46]

Indeed, the Society's *Transactions* during the period of Jefferson's life bears out the Society's heavy focus on natural history. Many of the Society's most prominent figures were thus not only Jeffersonian, but also the leading natural historians in America whose experimental endeavors studied the natural world, including Jefferson himself. The content thus represented the scientific agenda of APS Meetings and likewise the particular interests of its members who determined the papers to select. During the time of Jefferson's APS membership, 41 percent of all articles published by the Society from 1786 to 1825 focused on the field, far more than any other subject (Figure 2). The study of natural history became more prominent in works published by the Society during the 1790s, when Jefferson was most active in the Society and whose own greatest scientific contributions were in the field, growing from 38 percent of the Society's publications in 1786 to 51 percent in 1799. During Jefferson's stay in Philadelphia, then, what began as Jefferson's and the APS's apolitical interest in natural history, became a partisan issue as Jefferson became a symbol of one party, and the APS, of which Jefferson was so closely associated, served as an unwitting proxy for partisan attacks.[47]

It is, then, no coincidence that Jefferson, an outsider to the Federalist administrations that he served as secretary of state and vice president, found a more permanent home in the APS, an institution whose members seemed to have supported his politics alongside his science. Although Jefferson was unable to control the

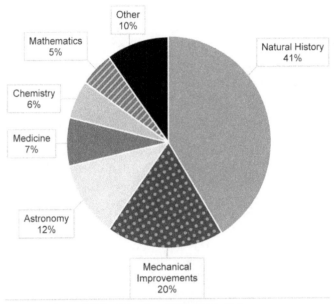

Figure 2. Topics published by the APS, 1786–1825.

government during these Federalist administrations, and indeed feared that the government undermined the very nation that he helped to create, the APS became the institution through which Jefferson could exercise some influence to advance his vision of a country with boundless potential that need not depend on government to realize it. He spent his years in Philadelphia helping the Society to establish its foundation and build its agenda, and he was joined in no small part by like-minded peers at the APS whose shared interests helped the Society succeed. By the time Jefferson departed Philadelphia, the APS had become a Jeffersonian institution in its approach to supporting science and its nationalistic research agenda, showing how science and politics could become interwoven in the Early Republic.

After assuming the presidency of the United States, Jefferson tried to decline reappointment to the APS presidency twice, believing that "the duties of another office" rendered him unable to perform the work that was required of him at the APS. He was so self-conscious of his absence from regular Society business that he began to apologize regularly to members, closing one letter to John Vaughan by noting that he hoped that his ability to acquire books on behalf of the Society would make him "good for something to the society." APS members continued to re-elect him, however, and Jefferson would remain the Society president until the end of 1814, when the Society finally accepted his resignation and appointed his close friend Caspar Wistar as president. Although Jefferson clearly felt guilty about his distance from the Society, Jefferson's ties to the APS, though changed by circumstance, remained strong through his term as president of the United States and especially during his retirement.[48]

Jefferson's most well-known interaction with the APS as US president was to send Meriwether Lewis to the Society in 1803 to prepare for his transcontinental voyage—what would become the realization of Jefferson's earlier failed Michaux Expedition. Lewis spent his days meeting with APS members and planning the trip. Lewis departed Philadelphia with instruments, medicines, and additional training in surveying provided by APS members. Lewis, himself a noted natural historian, was elected to membership that same year. Lewis's trip to Philadelphia typified the ways in which Jefferson's connection to the Society remained strong even in his absence from Philadelphia, and the ways the Society could serve his agenda still. Indeed, Jefferson used the network of friends and supporters he had established in person in Philadelphia for a variety of purposes throughout his US presidency.

First, Jefferson continued to rely on the APS for access to new knowledge. He was in frequent correspondence with APS Librarian John Vaughan, who was then building the Society's Library into a major repository and was also one of the most well-connected individuals in Philadelphia. Notably, Jefferson's correspondence with Vaughan is infrequent in the 1790s when they are in close proximity, but the explosion of letters that followed Jefferson's departure from the city reflects the close relationship the two forged in Philadelphia. Vaughan may have been one of the most frequent APS members with whom Jefferson maintained a correspondence, but he was not the only one. Other prominent members with whom Jefferson corresponded frequently on scientific matters included such APS stalwarts as Robert Patterson, Caspar Wistar, Benjamin Smith Barton, and Charles Willson Peale. Jefferson had virtually no known letters to these contacts before his time in Philadelphia, showing just how he used his relationships at the Society to acquire important scientific information.[49]

Topics of correspondence ranged, but they often reflected the core focus of natural history that Jefferson helped establish at the APS in the 1790s. Members wrote to Jefferson about excavations and new discoveries of bones. Jefferson also received numerous reports on Native American cultures, especially linguistic data, a personal interest of Jefferson's as well as a priority of the APS's collecting policy. Navigation and exploration were also frequent topics. Finally, and perhaps not surprising, one of the most frequent subjects of discussion was the distribution of books and papers to his friends at the APS.

This exchange of knowledge within the APS network also served as a conduit through which APS members could advise Jefferson on important issues while he was president of the United States. In 1803, for instance, John Vaughan shared new research on the efficacy of vaccination that the Society had helped produce. Knowing that Jefferson supported inoculation, Vaughan asked him to use his office to advocate for the adoption of a smallpox vaccine. Jefferson had been a proponent of the procedure, and continued to be as President. In another case, APS member and noted mathematician Robert Patterson developed a new method of ciphering. Jefferson and Patterson spent a great deal of time tinkering with the cipher and even involved APS member and minister to France Robert Livingston in the project, all with the goal of creating a more secure method of exchanging diplomatic correspondence (however, it failed because it was too complicated).[50]

Jefferson also appointed trusted APS members to important government positions. Andrew Ellicott is a prime example of how a connection formed through the APS developed into a government appointment. Ellicott was a Pennsylvanian who came to prominence in the 1780s because of his work surveying the boundaries of Virginia and Pennsylvania alongside David Rittenhouse and APS member and close Jefferson friend John Page. Although Jefferson had likely heard of Ellicott before coming to Philadelphia, through the APS the two developed a close relationship. Jefferson had come to admire him so much that he advocated for Ellicott's appointment as Surveyor of Washington, DC, and, later, as president of the United States, nominated Ellicott for Surveyor General for the Northwest Territory (an appointment Ellicott turned down for personal reasons). Instead, the appointment went to Jared Mansfield, who would eventually become an APS member. Ellicott's refusal seemed to have had little influence on their relationship. The two maintained a frequent correspondence in Jefferson's first term, second in frequency perhaps only to Vaughan. Later, Ellicott served as one of the many APS members who met with Meriwether Lewis and advised him on surveying techniques.[51]

Ellicott was not the only APS member Jefferson tried to enlist to aid his administration. In 1805, Robert Patterson, a trusted confidant who Jefferson first encountered in Philadelphia at the APS, did accept Jefferson's offer to direct the US Mint based in Philadelphia. Likewise, when Jefferson wanted to support an expedition in Louisiana, he asked William Dunbar, someone he came to know primarily through his contributions to the APS. Jonathan Williams, another APS acquaintance, was appointed the first commandant of the United States Military Academy in West Point, and member Isaac Briggs, a former assistant to Ellicott, was appointed surveyor general of the Missouri Territory. Jefferson thus leveraged the abilities of APS members whom he trusted to support his own government initiatives.

Ellicott, Patterson, Dunbar, Williams, and Briggs are all examples of APS members who served Jefferson in official capacities, but another contact Jefferson made at the APS served a significant extra-governmental role. Samuel Harrison Smith, a printer with Jeffersonian sympathies, was elected to the Society in 1797 after presenting a prize-winning essay on the need for free public education in a republic at an APS Meeting. The APS embraced Smith as a member, soon making him secretary of the Society under Jefferson until 1800. Jefferson also bonded with Smith, who, like Jefferson, was a Republican with a clear skill with the quill. The APS's endorsement of Smith's Republican proposal for education, his quick adoption into the Society's life, and his friendship with Jefferson reflect the ways Jefferson shaped the APS's operation.

After being elected president of the United States in 1800, Jefferson convinced his new friend to move his newspaper to Washington, DC, to serve as the official mouthpiece of the Jefferson administration. Newspapers were a notoriously difficult businesses to sustain, so Jefferson made sure his journalistic—as well as scientific—ally received what the editors of the *Papers of Thomas Jefferson* describe as "lucrative government contracts." It is likely that the genesis of this relationship began at the APS, as no known correspondence between the two precedes 1800.[52]

As president of the United States, Jefferson also had to determine the status of and access to scientific data acquired by the government, and here, the APS's position and Jefferson's were nearly identical. When asked by a correspondent whether scientific data collected by federal officials should be made public, Jefferson responded with an answer that echoed the APS's own stated idea about the use of knowledge: "my own opinion is that government should by all means in their power deal out the materials of information to the public in order that it may be reflected back on themselves in the various forms into which public ingenuity may throw it."[53]

Now at the head of the federal government, rather than opposed to it as he had been when the Federalists controlled the presidency, Jefferson also had to confront the issue of government funding of scientific endeavors. In several cases, he broke with his previous opposition to the direct funding of independent research initiatives, believing them of vital national interest. Jefferson used government funds to advance the scientific agenda he supported at the APS in the 1790s through federal initiatives, such as a robust surveying regime and expeditions, of which Lewis and Clark are but the best known of many who charted riverways, made scientific observations, and conducted ethnographic fieldwork in the areas of the American interior still largely unknown to Americans. He even supported projects that might appear to be of less clear national interest, such as Peale's excavations of a mammoth in Newburgh, New York. At Peale's request, Jefferson offered Peale the use of a patent pump from a naval frigate as well as some army tents. Jefferson's support of Peale was the natural evolution of Jefferson's belief that natural history was tied to a nation-building project and would also realize one of the APS's collecting goals. Although he supported such research through the APS while in Philadelphia, as president, the discovery of a mammoth, he believed, was of such national significance that he diverted federal resources to aid in its recovery. Although Peale ultimately turned down Jefferson's offer because it arrived too late, he did receive a loan from the APS to undertake the project.[54]

The funding of Peale's project showed just how interwoven the APS's agenda could be with Jefferson's national agenda at a moment when he was president of both the country and the APS. During Jefferson's administration, then, the APS served science and the nation in just the way Jeffersonians envisioned it would during the Federalist administrations. As with Meriwether Lewis's visit to the Society before embarking on his transcontinental expedition and with Peale's archeological dig, it was a nongovernmental body that was in service of the nation's interest.

Jefferson's experience with the APS also had a direct influence on his thinking about the establishment of a new learned society at West Point, the military academy he founded as president. After appointing fellow APS member Jonathan Williams to serve as the first superintendent of the school, he encouraged Williams to create another learned society that modeled itself on the APS: the United States Military Philosophical Society. Founded by Williams and other officers of the Corps of Engineers in 1802, the Military Philosophical Society sought to promote military science, but also to promote science and "natural philosophy" more generally among the officer corps and enlisted men of the new Army. With the motto *Scientia in Bello Pax* ("By Science in War Peace is Produced"), this new society followed in the APS's footsteps as a technically private institution explicitly dedicated to serving public purposes. Jefferson not only approved of the Society's creation, but he agreed to serve as honorary patron of it, similar in role to the APS presidency.[55]

Jefferson's involvement in this other institution also sheds light on his own thinking about the APS and the construction of its membership. Williams, after launching the institution, saw the offer of membership as a potential way to establish the infant organization's legitimacy. At one point, he considered making all members of Congress members in his learned society, believing it could help the institution gain much-needed support within the government, but Jefferson strongly cautioned him against such an idea. He warned Williams that he should not treat membership so casually and told him that he should include only members of Congress whose credentials fit with the Society's aims. As he noted to Williams, the indiscriminate selection of all members of Congress would "gratify no particular member, nor lead him to feel any more interest in the institution than he does at present." A targeted membership of like-minded congressmen, "friends of science, or lovers of the military art," Jefferson suggested, "will be gratifying to them inasmuch as it is a selection, and inspire them with the desire of actively patronising it's [*sic*] interests." Such a strategic approach to the Military Philosophical Society's membership likely reflected Jefferson's own thinking about APS membership. Jefferson was clearly an advocate of a thoughtful selection of individuals whose interests would serve the agenda of the institution, something that certainly seemed to define the APS's constituency during Jefferson's time in Philadelphia, even if such decisions were more implicit than explicit in the historic record for that earlier period.[56]

The relationship between Jefferson and the APS was not a one-way street in which Jefferson exploited his ties to the APS to serve his needs as president. Jefferson also deployed his office to bolster the Society's reputation and provide aid in service of its mission. For instance, Jefferson promised members that he would use his position to acquire books from Europe for the Society's holdings. In 1802, he vowed to promote the APS to European peer societies, assuring John Vaughan that he would "write letters to the most convenient Consuls" and request that they send scientific publications to the APS. Jefferson also forwarded important reports he received from the field that he felt might be worthy of publication in the *Transactions*. In terms of building the APS membership, Jefferson continued to introduce individuals who impressed him to the APS, either as correspondents with the Society or as new members. Indeed, as president of both the United States and the Society, Jefferson could better serve as a central node in a scientific network that stretched across the Atlantic. For instance, in 1801, he served as a bridge between the Academy of Geneva and the APS. Society members were aware of Jefferson's importance to their institution. As Jonathan Williams, a descendant of Franklin and an active APS member, noted in a letter to Jefferson, he was simultaneously "the President of the Philosophical Society, and The Cheif [*sic*] Magistrate of the Union," a dual responsibility that could aid the Society as much as it did the president personally.[57]

A clear example of the way Jefferson's stature as president of the United States aided the Society came in 1805, when Jefferson received a letter from a French correspondent reporting on recent experiments conducted there on navigation. The Frenchman asked Jefferson to encourage the APS to pursue similar experiments in the American context. Jefferson appeared to agree with the recommendation and forwarded the correspondence to Caspar Wistar at the APS. The Society then debated the proposal and formed a committee to evaluate the viability of the research. The committee agreed that the topic was worthy of further investigation and voted to publish a call for citizens to provide data needed to conduct the work. As this and several other similar cases demonstrate, even though Jefferson had less time to commit to the Society, he still influenced the APS's agenda.[58]

More important, as president, Jefferson used his office to help connect the Society with peers in Europe and throughout the American continent. In fact, during his lifetime, international members of the APS formed close to 40 percent of all APS members, significantly more than the approximately 15 percent they constitute today. The large number of international and primarily European members represented the APS's desire to connect with leading minds abroad, many of whom were producing the most cutting-edge new research. It also needed a large foreign contingent as a way to establish its credentials and legitimacy in the eyes of those in Europe who could confer such status. Jefferson's new role as president of the United States thus allowed him to better raise the APS's reputation amid its European peers by serving as a key connector, even if he was less directly involved in the organization's day-to-day life.[59]

If "another office" had distracted Jefferson from his scientific
pursuits while president of the United States, in retirement,
Jefferson's connection to the APS grew stronger even if he
remained physically distant from it. In fact, in 1810, after receiving
word of his re-election as president of the Society, instead of
turning the post down as he had earlier, he embraced the position,
writing that he would work to "forward their laudable pursuits
for the information & benefits of mankind." The Society, too,
recognized that his retirement might allow him more time to
dedicate to APS business and the nobler pursuit of knowledge.
"Retiring from the direction of public affairs," the Society's secretary wrote to him after his re-election, "the Philosophic Patriot
posesses [*sic*] a usefulness and enjoys a happiness unknown to
the mere Statesman."[60]

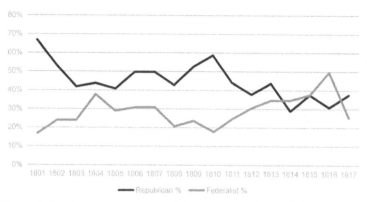

Figure 3. Partisan composition of the elected officers of the APS, 1801–1817.

By the time of Jefferson's retirement in 1809, the Society itself was undergoing its own transition. Although the officers of the Society remained largely Republican throughout Jefferson's presidency of the Society, the partisan divide of the Society that marked the hyper-partisan 1790s waned as the capital moved to Washington, DC. Indeed, the Society's leadership became more bipartisan in the less politically charged 1810s, as exemplified by the close friendship of two of the most prominent figures in the operation of the APS, Republican and Jefferson protégé Peter Du Ponceau and Federalist and close Jefferson friend John Vaughan. By the 1810s, in contrast to the 1790s, the Society's leadership positions showed no discernible difference between the number of Federalists and Republicans elected to office (Figure 3).

Jefferson nonetheless remained an active member of the Society, and patterns established in Philadelphia decades before continued in his retirement. His relationship with Vaughan grew deeper even if they were separated by distance and party. Indeed, from 1816 until Jefferson's death in 1826, Vaughan was Jefferson's ninth most frequent correspondent. By 1818, Vaughan served as Jefferson's conduit for wine and books from Europe. Jefferson's correspondence with APS members expanded to include new members and leaders of the Society whom he likely never knew except by letter. The content of these exchanges remained, as they always had, focused on new research and discoveries, often dealing with natural history.[61]

Jefferson had always provided materials for the Society, but in retirement, he devoted more time to acquiring items for the Library. Jefferson, the president during the Society's adoption of its first collecting policy, became particularly involved when the Society created a new "Historical and Literary Committee" in 1815 to oversee a renewed effort to increase acquisitions. The committee concluded that, as the APS grew alongside the nation, its collecting should evolve to include important historical manuscripts and artifacts that related to the country's political history. "The Communications of patriotic & literary characters in every part of the Union," Peter Du Ponceau explained in a letter he sent to Jefferson outlining this decision, "will be of great use to the future historians." Though this represented a change to the Society's focus on science, especially natural history, Jefferson, a central figure in this history, was an enthusiastic proponent of the expanded mission.[62]

Between 1815 and 1818, Jefferson corresponded extensively with Society members, especially Du Ponceau, about the work of the committee. He also followed his words of support with action. He gave of his own material and acquired several valuable manuscripts from friends. His most notable contributions were the few remaining remnants of his Indian vocabularies, William Byrd's work on drawing the line between Virginia and Carolina, and the journals of Lewis and Clark. Indeed, although the APS had served as a refuge for Jefferson to escape the partisan sniping in Philadelphia, it now served as an outlet through which Jefferson could engage in the life of the mind from his refuge in Monticello.[63]

As Jefferson became more ensconced at Monticello, he yearned for a source of intellectual stimulus that would rival that which he had experienced in Philadelphia. He decided that Virginia needed a place of higher education open to all, a university he called an *academical village*. Beginning in the 1810s, but picking up speed in 1818, Jefferson dedicated much of his later life to founding this institution of higher education, the University of Virginia (UVA), in his home state. A central piece of this place of learning was, of course, its library. A bibliophile with extensive experience contributing to libraries like the APS, Jefferson took a direct role in building a collection at UVA of international import. Such energy focused in his backyard, however, came at the expense of his interest in the Society's book collection. The records show that before 1819, Jefferson was a regular contributor of books, materials, and advice to the APS. After UVA was founded, however, Jefferson's donation of materials to the APS Library ceased.

One other potential reason for Jefferson's reduced involvement in the APS's collection was the private sale of his library to the federal government in 1815, which followed the British burning of Washington during the War of 1812. Indeed, that decision engendered what was perhaps Jefferson's only contentious interaction with the Society. As Jonathan Williams told Jefferson, Jefferson's disposal of books surprised some APS members, who had expected that Jefferson would contribute his greatest scientific volumes to the Society's growing Library. "Works of History, Law, Government Finance, political Economy, and general Information, may with propriety be so deposited," Williams conceded, "but such Books as would adorn our Library and aid this Society in 'the promotion of useful knowledge' must there become motheaten upon the Shelves." Jefferson never responded, but his divestment of much of his collection to Congress alongside his devotion to UVA's library meant that he had less to give to the APS.[64]

But even if Jefferson's engagement with the APS waned as he aged and he instead focused his intellectual energy on building an "academical village" in Virginia, the Society's influence was very much alive on the campus of his university. Its aim, after all, was to do for the masses what the APS did for the elite. As the University Rectors wrote:

We fondly hope that the instruction which may flow from this institution, kindly cherished, by advancing the minds of our youth with the growing science of the times, and elevating the views of our citizens generally to the practice of the social duties and the functions of self-government, may ensure to our country the reputation, the safety and prosperity, and all the other blessings which experience proves to result from the cultivation and improvement of the general mind.[65]

If his willingness to contribute to the APS Library diminished later in life, his involvement with other parts of the APS's functions nonetheless remained active, although certainly on a much smaller scale than earlier in his life. In 1818, he was elected to a three-year term as a Councilor, and re-elected in 1821 and 1824. Though he never attended a Meeting or took much active part in Society affairs after his correspondence with the Historical and Literary Committee ended in early 1819, he continued to write regularly to Vaughan, Peale, and other APS members, though the volume of correspondence decreased.

Indeed, if one of the primary ways that the APS hoped to advance knowledge was to use membership as a means of creating a network of correspondents, then Jefferson's lifetime commitment and connection to the APS via written letters becomes even more pronounced. A survey of his correspondence suggests that perhaps as much as 26 percent of Jefferson's lifetime writing was carried out with APS members. Though this total includes close political allies like Madison, Gallatin, and Adams, once these more political affiliates are excluded, the total is still at 11 percent, a significant sum (nearly 5,000 letters). Within this correspondence, Jefferson engaged on a personal level with many different APS members. Our database reveals that Jefferson exchanged at least one letter with over 34 percent of those who served as resident members during his lifetime (222 different members), as well as 22 percent of the international members (67 total). With most (72 percent) of these letters, at least some of the correspondence concerned science, scholarship, or the APS itself.

And although Jefferson's involvement with the APS tapered as he aged, it never completely stopped. He renewed contact with his close friend Robert Patterson after a multiyear hiatus, for instance, to turn over some responsibilities of APS membership to a younger generation. A request from a French scientist to open an exchange of ideas spurred Jefferson to write to Patterson in hopes that he might find a younger APS member to maintain a correspondence with the Frenchman. Arthritis and age, he told Patterson, had rendered him unfit for such intensive relationship-building.[66]

Still, even if Jefferson reduced his contact with the Society, the Society and its members nonetheless remained a part of his intellectual life until the end. Indeed, even though donations of materials to the APS Library stopped, he still supported the institution and its collection through other means. In 1824, with debt piling up, he found funds to give to the Society to help defray the costs of publishing the APS Library's first catalog, a major undertaking led by his close friend and APS Librarian John Vaughan. Two years later, in his last letter to Vaughan written less than three months before his death, he described Vaughan in what was clearly a farewell note as "a friend of another century," a clear and touching reference to the friendship that Jefferson realized was forged in Philadelphia and sustained by their joint involvement in the APS.[67]

Examining Jefferson's long and deep relationship with the APS reveals that the institution played a significant role in Jefferson's life and, conversely, Jefferson played a key role in the Society's development as it evolved in the Early Republic. It's fair to say that if Benjamin Franklin is credited with having founded the APS in 1743, then it should also be said that Jefferson built the house that sits on Franklin's foundation. Jefferson contributed to nearly every program of the Society as the organization was reestablishing itself in the wake of the American Revolution. He helped establish the *Transactions* as one of the leading scientific journals in the world by publishing an article that served as a landmark in the history of American science and encouraging others to share their research through the publication. When in Philadelphia, he attended nearly every Meeting. When away from the city, he forwarded research findings that he felt might make for important topics for discussion. He nominated new members, giving shape to the intellectual community that formed the core of the Society. He used his connections in Europe to build ties between the APS and peer organizations across the Atlantic, and he leveraged the power of his office to acquire materials from abroad for the Society. He also provided guidance to the Society's collecting policy and actively contributed to it. His long engagement with the collections helped create the three core collecting areas of Native American history and culture, the history of science, and early American history that form the corpus of the 13 million pages of manuscripts held by the Society today.

While Jefferson had a lasting influence on the Society—one that still guides significant aspects of its programming today—an examination of the records also shows that the Society gave much to Jefferson. Probably more than any other single institution, the Society provided Jefferson with intellectual sustenance. In Philadelphia, while secretary of state and vice president in Federalist administrations, Society Meetings in which the latest discoveries were discussed and debated served as a sanctuary for Jefferson amid the charged partisan environment of the 1790s. Indeed, Jefferson found in the Society a means through which he could simultaneously pursue his own scholarly agenda while also fulfilling parts of his political agenda (and often the two went hand in hand). After leaving Philadelphia, the Society continued to be a venue through which Jefferson pursued his scientific interests because of the correspondence network he had established through his ties to the APS. In retirement, the APS's network that Jefferson first forged in Philadelphia became even more important to him. Cloistered away at Monticello, his connection to the APS provided him with a way to stay engaged with the world of ideas occurring far from the hilltop upon which he resided.

In short, the APS simultaneously served and fueled Jefferson's intellectual development and scholarly pursuits. Jefferson, for his part, used his power to establish the Society as the central node from which early American science would be conducted. He did so in no small part because during his time in Philadelphia, the APS had become a Jeffersonian institution whose members shared his agenda and were loyal to him, a legacy that continued well into the nineteenth century and, indeed, still shapes the institution today.

Afterword:

Looking Back and Looking Ahead

I was flattered when I learned that the American Philosophical Society wanted to reissue my short piece "The Other Presidency: Thomas Jefferson and the American Philosophical Society." I was also a little surprised. The essay was written as part of the 275th celebration of the APS's founding in 1743. It was printed as a small pamphlet in 2018 and then more formally as an article in the APS's peer-reviewed journal *Proceedings of the American Philosophical Society*. Given this history, when I was first approached about publishing it as a book for the APS Press, I initially had a hard time seeing it as "a book." That certainly was not its original conception. But, after conversations with Peter Dougherty at the APS Press, I realized that maybe it did have the potential to be a monograph. At the very least, reissuing this piece as a standalone volume might engender wider distribution by providing another means of access—and, in so doing, continue to contribute to conversations about Thomas Jefferson's leadership and legacies across several arenas.

Although the impetus for publishing this piece originally was the Society's anniversary, the real spur for me personally was an observation about Jeffersonian scholarship. There is, to be sure, a surfeit of work on Jefferson. He is considered the most enigmatic, the most complicated, and, arguably, the most complex of the so-called Founding Fathers of the United States. He is, as Joseph Ellis dubbed him, "the American sphinx." There are books on Jefferson and religion, on Jefferson and science, on Jefferson's intellectual development, on Jefferson's political philosophy and on his political economy, on Jefferson as a writer, and, of course, a regular churn of biographies that all claim to shed new light on or offer a more complete picture of the man. This steady stream of Jeffersonian scholarship suggests that people remain hungry to learn more about the individual who was simultaneously both our nation's and the Society's third president.

Yet, as the Librarian of the American Philosophical Society, a position that is steeped in the APS's institutional history, I was left underwhelmed when I turned to these books to learn more about the role that the Society played in Jefferson's own life—as well as what role Jefferson played in the life of the APS. In most of these works, there was no mention of the APS at all. If there was, it was often just a passing reference, noting Jefferson's membership and presidency. Yet, Jefferson served as president of the Society for seventeen years, from 1797 to 1814. Surely, I thought, there must be something more to say. As it turns out, there was. The APS itself, including another of its former Librarians, Ted Carter, had published several pieces on Jefferson and the APS, most of which I believe I cited in my piece. My humble intervention, aided by research assistance by Abigail Shelton and John

Kenney, PhD, was to pursue that same line of inquiry by using new digital tools to add more to this scholarship and, maybe, shape future scholarship on Jefferson.

In terms of the research for the article, one of my goals was to take advantage of the Jefferson Papers project, especially its digital versions, to document Jefferson's correspondence with APS members and with the institution itself. These digital databases, along with the invaluable research assistance from John Kenney and Abigail Shelton, made it possible to better quantify Jefferson's tangible connections to the Society and vice versa. It is clear, to me at least, that Jefferson's activity at the APS enabled him to forge important bonds with fellow members, people like Caspar Wistar, Robert Patterson, and John Vaughan, that lasted far longer than Jefferson's stay in Philadelphia. Perhaps Jefferson would have made these same friendships had the APS not existed to bring them together around their common interests, but its presence, and the regular meetings Jefferson attended with these colleagues, certainly helped strengthen them.

Another contribution, in my estimation, was to show how active Jefferson was in the affairs of the Society, and how the Society actively supported Jefferson's own intellectual pursuits. When Jefferson was stationed in Philadelphia as the first secretary of state, he also became a chief driver of the Society's agenda, most notably by serving on the Historical and Literary Committee, the committee that essentially laid the foundation for the Society's distinguished library. If Jefferson shaped the institution's direction and holdings, the Society also directly encouraged Jefferson's own pursuits. The most notable example of this support was the APS's underwriting of Jefferson's expedition with James Madison to study the Hessian fly in the northern states, an endeavor that, coming in 1791, just as partisan feelings were setting in between Federalists led by Alexander Hamilton and Republicans led by Jefferson and Madison, also had significant political overtones.

In fact, one of the findings that became apparent the more I worked on the project was that while the APS was primarily a scientific organization, it also became politicized in the Early Republic. It was subtle, to be sure, but there is no doubt, as I hope the essay makes clear, that the APS had become a thoroughly Jeffersonian institution, both in terms of its scientific agenda and its political affiliations.

As I hope is obvious, I am still excited by these findings, and about the possibilities for future research along these lines. Part of my surprise when Peter asked me to write a retrospective conclusion on this piece was that it simply does not feel that old to me; five years did not seem long enough to look back at a piece that still felt fresh. And yet, as I re-read it, and as I thought about the other things that have been uncovered even in the five years since its publication, the more I realized what I had to say was just the tip of the iceberg.

The APS is a dynamic research center, filled with fellows and researchers making discoveries in its collections every day. Its staff, too, are engines for scholarship, and since this essay was published, a number of projects have helped the APS better understand its early history and Jefferson's role in it.

First, there are projects underway to analyze the early catalogs of the APS Library, with the hope that we may be able to digitize and transcribe them so other scholars can conduct their own separate analysis of this vital early research collection. This project is likely going to reveal in even greater depth Jefferson's continued involvement with the Society. I know that just by thumbing through the pages of these records, Jefferson's acts of generosity are regularly revealed in the provenance notes of many books. Related to that, Jefferson's involvement in the Society's Historical and Literary Committee—really its first collections committee—is an underappreciated aspect of Jefferson's intellectual life, and further research into both his involvement in it and its acquisitions can shed new light on the collecting priorities not just of the APS but of Jefferson and the scientific community of the young nation.

Staff have also worked on quantifying the geographic make-up of the Society's earliest members with a particular eye toward its foreign members. Unsurprising, the data reveals a fairly sizable percentage hailing from Europe, especially France and Great Britain. Further analysis will reveal, I suspect, the hand of both Jefferson and Franklin in the nomination of many of these international members, and especially Jefferson's in those elected from France in the 1790s and beyond. These international connections, and the ways that intellectual ties might have intersected with political, diplomatic, and economic aims in the Early Republic is another potentially fruitful avenue of further study.

Finally, there are also many stories, mentioned as evidence in the essay, that could benefit from further research, analysis, and explication. Episodes like the Society's support of the Michaux Subscription List, and Jefferson's leading role in its planning, can tell us much about the priorities of the institution and, though it, the nation. Jefferson and Madison's trip to investigate the Hessian fly has been told often, but the APS's involvement in it has been less examined. In short, there is simply much more to be told, and many of these stories will be of interest to both the general public and to scholars. APS staff and researchers are regularly conducting research in and on the Society's collections, and I suspect that as they do more, we will learn even more about these fascinating and important stories that connect Jefferson, the APS, science, and politics.

In short, my sincere wish is that this essay proves only a starting point for future scholarship. I remain convinced that there is even more that can be done to show the importance of the APS to Jefferson's life and, just as important, how important the APS was to Jefferson—and, by extension, how significant the

APS was as a cultural and intellectual institution in the Early Republic and beyond. Continued work along these lines will, I believe, highlight this key aspect of the APS's early history, one known to those familiar with the APS but not much further beyond it: the American Philosophical Society played a central role in cohering and supporting the United States' first national scientific community as well as the nation's involvement in international scientific conversations. It was (as it still is) a dynamic institution that deserves far greater prominence in the history books than it tends to receive.

Such work does pose a challenge, though, one that I dealt with when first writing this essay and one that might explain why these lines of inquiry into Jefferson's life and the APS's history have largely gone unpursued. One worries that such research might produce work that others deem too much of a hagiography, either of an individual or an institution, or it may come across as bland, self-congratulatory institutional history. There is, perhaps, a way to avoid these traps. Rather than only myopically focusing on an individual or an institution, scholars can situate either Jefferson or the APS within their wider cultural and social contexts, to use them, in some way, as windows into the world in which they operated and influenced the course of events. Doing so, I suspect, may provide even more insight into the complexities of this revolutionary age, one in which both Jefferson and the APS played prominent roles.

Timeline of the Other Presidency

This timeline represents a sampling of Thomas Jefferson's contributions to the APS, contextualized with reference to the major events of Jefferson's life.

1780: Thomas Jefferson is elected to the APS.

January 3, 1783: Jefferson attends a Meeting and presents the works of the Italian natural philosopher Abbé Fontana.

January 8, 1783: Jefferson attends a Meeting and motions that the APS commission an orrery to be made by David Rittenhouse and presented to the king of France.

August 1784: Jefferson arrives in France, where he takes up the position of American minister, replacing Benjamin Franklin.

1785: Jefferson publishes his *Notes on the State of Virginia*, which he had written between 1781 and 1783 at the behest of the French minister, François Barbé-Marbois, an APS member.

June 1787: Jefferson sends transactions from the Royal Academy of Sciences in Turin to the APS.

September 26, 1789: Jefferson departs France.

March 1790: Jefferson takes up President Washington's appointment as secretary of state.

December 6, 1790: The US government moves from New York to Philadelphia, where it will remain for the next decade.

April 15, 1791: Jefferson attends a Meeting and motions that a committee be formed to collect materials for a natural history and prevention of the Hessian fly. Jefferson is appointed to the committee.

May 1791: Jefferson forwards Transactions from the Royal Academy of Sciences in Paris that he received from Nicholsa de Condorcet.

August 19, 1791: Jefferson presents a piece of Native American sculpture that he received from Henry Innes and a paper from J. Cooke on new standards for weights and measures.

March 15, 1793: Jefferson presents a letter and volume donated by a European contact, Rudolph Vall Travers.

1793: Jefferson forwards an essay from Dr. Benjamin Rush on the sugar maple to the APS for publication.

December 31, 1793: Jefferson resigns as secretary of state and returns to Virginia.

May 29, 1795: Jefferson arranges for the APS to exchange specimens with the Leverian Museum in England through Rudolph Vall Travers.

January 6, 1797: The APS elects Jefferson as Society president.

March 4, 1797: Jefferson is inaugurated as second vice president of the United States in Philadelphia.

March 10, 1797: Jefferson presents bones and a paper on the Megalonyx.

May 19, 1797: Jefferson presents two books from W. Valtravers and motions to form a committee for collecting the antiquities of North America.

January 5, 1798: Jefferson presents the Society with a Swedish coin given to him by Polish general and veteran of the American Revolution, Thaddeus Kościuszko, and a mammoth bone found in Virginia.

May 20, 1798: Jefferson presents a paper by T. C. Martin on a new hand-threshing machine and presents his own paper on a new design for a mouldboard plow.

February 25, 1799: Jefferson connects Charles Willson Peale with Louis of Parma to exchange specimens.

1799: The 1799 *Transactions* publish several articles forwarded by Jefferson, including his own contributions on the Megalonyx and new plow design.

1800: The seat of government of the United States moves from Philadelphia to Washington, DC.

December 1800: Jefferson forwards a prospectus of the Royal Institution that he received from the American-borne Sir Benjamin Thompson, Count Rumford, to the APS.

December 1800: Jefferson donates a paper by Hugh Williamson on the climate in Quebec and Benjamin Waterhouse's *A Prospect for Exterminating the Small-pox.*

March 4, 1801: Jefferson is inaugurated as third president of the United States in Washington, DC. He appoints several APS members to senior positions—James Madison as secretary of state, Albert Gallatin as secretary of the Treasury, and Robert R. Livingston as minister to France.

1801: Jefferson offers APS member Andrew Ellicott the position of surveyor general of the Northwest Territory. Ellicott declines the offer.

December 1801: Jefferson donates a box of specimens from Jose Garcia Armenteros (secretary of the Royal Philippine Company) and a copy of Andrew Ellicott's astronomical observations.

December 1801: Jefferson appoints Jonathan Williams, an APS member, as the first superintendent of West Point.

January 1802: Despite an attempt to decline the honor, Jefferson is re-elected as president of the APS. He will continue to be re-elected for the next 12 years.

December 25, 1802: Jefferson agrees to become patron of the United States Military Philosophical Society, founded by West Point Superintendent Jonathan Williams, an APS member.

May 14, 1804: The Lewis and Clark Expedition, established by Jefferson to explore the new Louisiana territory, sets out.

1805: Jefferson appoints APS member Robert Patterson as director of the US Mint.

November 15, 1805: The APS receives a donation through Jefferson from Captain Meriwether Lewis of a box of plants, earth, and minerals.

December 20, 1805: Jefferson's donation of a horned lizard from upper Louisiana is presented at an APS Meeting.

July 18, 1806: The APS receives a donation of 150 Roman bronze coins from Jefferson. Jefferson received these coins from N. H. Weinwich, the secretary of the Royal Society Heraldry and Genealogy in Denmark.

February 6, 1807: Jefferson sends the Society the head of a marmot discovered in Virginia by Colonel John Stewart.

March 4, 1809: Jefferson's second term as president of the United States comes to an end.

1809: In the 1809 volume of the APS *Transactions*, the following donations from Jefferson are recorded:

- Documents relative to the late discoveries in exploring the Missouri, Red, and Washita Rivers (1806, 8 vols.)

- Documents relative to Louisiana

- *Description of the Sonde de Mer, ou Batometre* by A. Van S. Luiscius (1804, 4 vols., with model)

- *Traite des moyens de deinfecter l'air de prevenir la contagion* ... by L. B. Guyton Morveau (1805, 8 vols.)

- *Histoire de laeur introduction dans les divers Etats d'Europe, et au Cap de Bonne Esperance* (1802, 8 vols.)

- *Du Cottonier de de sa culture* (1808, 8 vols.)

- *Sur les plantations des cannes a sucre en France*

- *Notes on the State of Virginia* (8 vols.)

- Also in the 1809 *Transactions*, several articles sent by Jefferson are published:

  - Extracts from a Letter, from William Dunbar Esq. of the Natchez, William Dunbar, 1809

  - "On the Language of Signs among Certain North American Indians," William Dunbar, 1809

  - "Meteorological Observations, for One Entire Year. Made by William Dunbar, Esq. at the Forest Four and a Half Miles East of the River Mississippi ..." William Dunbar, 1809

  - "Description of a Singular Phenomenon Seen at Baton Rouge," William Dunbar, 1809

November 23, 1814: Jefferson at last retires from the presidency of the APS and is replaced by Caspar Wistar.

February 1815: Jefferson sells his personal library to Congress, forming the basis of the modern Library of Congress.

January 22, 1816: Jefferson donates Benjamin Hawkins's "A Sketch of the Creek Country, in the years 1798 and 1799" to the Historical and Literary Committee.

December 1817: Jefferson sends the Historical and Literary Committee a manuscript copy of William Byrd's "Secret History of the Dividing Line."

January 1818: Jefferson is elected to a three-year term as a Councilor of the APS.

1818: The following donations from Jefferson were recorded in the 1818 volume of the *Transactions*:

- ○ Two skeleton heads found in the Big Bone Lick

- ○ MS memoir on the boundaries of Louisiana, many Indian vocabularies with a digest of several by himself

- ○ Lewis and Clark's original manuscripts

January 25, 1819: Jefferson's University of Virginia, in Charlottesville, is granted a charter by the Commonwealth of Virginia. Donations of material from Jefferson to the APS cease.

January 1821: Jefferson is re-elected to another three-year term as a Councilor of the APS.

January 1824: Jefferson is re-elected for another three-year term as a Councilor of the APS.

July 4, 1826: Jefferson dies at home in Monticello.

Notes

1. The two most complete works on Jefferson and the offices he held are Gilbert Chinard, "Jefferson and the American Philosophical Society," *Proceedings of the American Philosophical Society* 87, no. 3 (1943): 263–76; and Edward C. Carter, "Jefferson's American Philosophical Society Leadership and Heritage," in *"The Most Flattering Incident of My Life": Essays Celebrating the Bicentennial of Thomas Jefferson's American Philosophical Society Presidency, 1797–1814* (Philadelphia: Friends of the American Philosophical Society Library, 1997), 9–15.

2. Thanks to Google Books, mentions of the APS can be more easily quantified. In total, Peterson mentions the APS sixteen times. But if one excludes its inclusion in the bibliography and index, the APS appears in the text only eleven times, and of those the APS is just of passing note. See Merrill D. Peterson, *Thomas Jefferson and the New Nation: A Biography* (New York: Oxford University Press, 1970).

3. Joseph J. Ellis, *American Sphinx: The Character of Thomas Jefferson* (New York: Vintage, 1996).

4. Jon Meacham, *Thomas Jefferson: The Art of Power* (New York: Random House, 2012).

5. John B. Boles, *Jefferson: Architect of American Liberty* (New York: Basic Books, 2017).

6. Kevin J. Hayes, *The Road to Monticello: The Life and Mind of Thomas Jefferson* (New York: Oxford University Press, 2008).

7. Silvio A. Bedini, *Thomas Jefferson: Statesman of Science* (New York: Macmillan, 1990).

8. Keith Thomson, *Jefferson's Shadow: The Story of His Science* (New Haven, CT: Yale University Press, 2012).

9. Chinard, "Jefferson and the American Philosophical Society," 263.

10. Carter, "Jefferson's American Philosophical Society," 9–15, especially p. 10. Bedini also noted the meteorological diary in *Statesman of Science*, 84.

11. As Gilbert Chinard noted, the printed version of the Minutes are incomplete and suffer from a heavy editorial hand that excised potentially pertinent information. Chinard, "Jefferson and the American Philosophical Society," 263. The APS Library recently digitized the early Minutes of the American Philosophical Society until 1842. See: http://diglib.amphilsoc.org/islandora/graphics/minutes-american-philosophical-society.

12. Benjamin Franklin, *A Proposal for Promoting Useful Knowledge among the British Plantations in America, May 14, 1743* (Philadelphia, 1743).

13. For the best study of this new group, see Whitfield J. Bell, Jr., "History of the Society," in *Patriot-Improvers: Biographical Sketches of Members of the American Philosophical Society*, vol. 1 (Philadelphia: American Philosophical Society, 1997), 339–46, quote on 341.

14. Bell, *Patriot-Improvers*, 341–45.

15. Bell, *Patriot-Improvers*, 345 for merger; and quote from W. Lane Verlenden, "The American Society Held at Philadelphia

for Promoting Useful Knowledge," *The Pennsylvania Magazine of History and Biography* 24, no. 1 (1900): 1–16.

16. For more on Rittenhouse, see Brooke Hindle, *David Rittenhouse* (Princeton, NJ: Princeton University Press, 1964), especially chapter 4, "The Transit of Venus"; see also "John Page to Thomas Jefferson," April 23, 1785, in *The Papers of Thomas Jefferson: Digital Edition*, eds. James P. McClure and J. Jefferson Looney (Charlottesville: University of Virginia Press, Rotunda, 2008–2018).

17. For more on the Silk Society and its purpose, see Zara Anishanslin, "Producing Empire: The British Empire in Theory and Practice," in *The World of the Revolutionary American Republic: Land, Labor, and the Conflict for a Continent*, ed. Andrew Shankman (Routledge, 2014), 27–53, especially 42–45.

18. Bell discusses the unwinding of the project in "Charles Thomson," in *Patriot-Improvers*, 184. For a list of funders, see *Directions for the Breeding and Management of Silkworms* (Philadelphia: Crukshank and Collins, 1770), xi–xv.

19. Quote from "The Virginian Society for the Promotion of Usefull Knowledge," *Colonial Williamsburg Journal*, Autumn 2003, http://www.history.org/foundation/journal/autumn03/society.cfm. Note that the title of the organization was never stable, but I have adopted in the text the title used in the advertisement announcing its creation, a title also used by Hayes and Bedini in their respective works.

20. For more on the Virginia Society and Jefferson's involvement in it, see "Memorandum Books, 1773," *Founders Online*, National Archives, last modified June 13, 2018, http://founders.archives.gov/documents/Jefferson/02-01-02-0007, especially fn. 42 [Original source: James A. Bear, Jr. and Lucia C. Stanton, eds., *The Papers of Thomas Jefferson*, Second Series, *Jefferson's*

Memorandum Books, vol. 1 (Princeton, NJ: Princeton University Press, 1997), 301–54]; and "John Page to Thomas Jefferson," for its end. See also Hayes, *Road to Monticello*, 129, 140; Bedini, *Statesman of Science*, 59; and "The Virginian Society." For the fire, see "John Page to Thomas Jefferson."

21. Edward C. Carter, *One Grand Pursuit: A Brief History of the American Philosophical Society's First 250 Years* (Philadelphia: American Philosophical Society, 1993).

22. Carter, "Jefferson's American Philosophical Society," 10.

23. For Jefferson's science, see Thomson, *Jefferson's Shadow*; see also Bedini, *Statesman of Science*. For Jefferson's gardening pursuits, see Andrea Wulf, *Founding Gardeners: The Revolutionary Generation, Nature, and the Shaping of the American Nation* (New York: Vintage, 2012), especially 173–89; see also Peter Loewer, *Jefferson's Garden* (Mechanicsburg, PA: Stackpole Books, 2004), 7. For an example of an exchange between Jefferson and an APS Member for seeds, see Thomas Jefferson to John Bartram, Paris, January 27, 1786, "From Thomas Jefferson to John Bartram, with Enclosure, 27 January 1786," *Founders Online*, National Archives, last modified April 12, 2018, http://founders.archives.gov/documents/Jefferson/01-09-02-0201 [Original source: Julian P. Boyd, ed., *The Papers of Thomas Jefferson*, vol. 9, *1 November 1785–22 June 1786* (Princeton, NJ: Princeton University Press, 1954), 228–30]; see also "Thomas Jefferson to John Bartram, Jr.," June 11, 1801, in *The Papers of Thomas Jefferson*, ed. Barbara Oberg (Princeton, NJ: Princeton University Press, 2007), 34, 306. For detailed information on Jefferson's meteorological diary, see Lucia Stanton, "Weather Observations," *Thomas Jefferson Encyclopedia*, accessed October 14, 2018, https://www.monticello.org/site/research-and-collections/weather-observations. For a close study of Jefferson's linguistic work, see Peter Thompson, "'Judicious Neology': The Imperative of Paternalism in Thomas Jefferson's Linguistic Studies," *Early American Studies* 1, no. 2 (2003): 187–224.

24. "APS Minutes, 1787–1793," *APS Library*, accessed May 5, 2018, http://diglib.amphilsoc.org/islandora/object/american-philosophical-society-minutes-1787-1793; see also "APS Minutes, 1793–1798," *APS Library*, accessed May 5, 2018, http://diglib.am philsoc.org/islandora/object/american-philosophical-society-minutes-1793-1798#page/1/mode/1up.

25. The difference in these views is well established in the historiography. The best summary of this divide likely remains Stanley Elkins and Eric McKitrick, *The Age of Federalism: The Early American Republic, 1788–1800* (New York: Oxford University Press, 1993).

26. See Elkins and McKitrick, *Age of Federalism*, 262–63, 279–80; Andrew Burstein and Nancy Isenberg, *Madison and Hamilton* (New York: Random House, 2013), 236; and Noble E. Cunningham, Jr., *Jefferson vs. Hamilton: Confrontations that Shaped a Nation* (New York: Macmillan, 2000), 75, for the Jeffersonian critique of manufacturing. For quote, see Thomas Jefferson, "Notes on the Constitutionality of Bounties to Encourage Manufacturing," February 1792, in *Papers of Thomas Jefferson*, ed. Charles Cullen, 23:172–73.

27. See Wulf, *Founding Gardeners*, 81–99; see also Brooke Hunter, "The Forgotten Legacy of the Hessian Fly," in *The Economy of Early America: Historical Perspectives and New Directions*, ed. Cathy Matson (University Park, PA: Penn State Press, 2006), 236–62.

28. For information on the committee and Jefferson's involvement, see "APS Minutes, 1787–1793"; "Notes on the Hessian Fly [1–15 June 1792]," *Founders Online*, National Archives, last modified April 12, 2018, http://founders.archives.gov/documents/Jefferson/01-24-02-0004 [Original source: John Catanzariti, ed., *The Papers of Thomas Jefferson*, vol. 24, *1 June–31 December 1792* (Princeton, NJ: Princeton University Press, 1990), 11–14]; and "II. Jefferson's Notes on the Hessian Fly [24 May–18 June 1791],"

Founders Online, National Archives, last modified April 12, 2018, http://founders.archives.gov/documents/Jefferson/01-20-02-0173-0003 [Original source: Julian P. Boyd, ed., *The Papers of Thomas Jefferson,* vol. 20, *1 April–4 August 1791* (Princeton, NJ: Princeton University Press, 1982), 456–62]. For an example of someone responding to the circular, see "Jonathan Havens and Sylvester Dering to Thomas Jefferson," November 1, 1791, in McClure and Looney, *Papers of Thomas Jefferson.*

29. Hunter, "Forgotten Legacy of the Hessian Fly," 243.

30. Linda K. Kerber succinctly describes this divide over science in Linda K. Kerber, *Federalists in Dissent: Imagery and Ideology in Jeffersonian America* (Ithaca, NY: Cornell University Press, 1970), 75.

31. Kerber, *Federalists in Dissent,* 74–76; see also Bedini, *Statesman of Science,* 260–62.

32. Kerber, *Federalists in Dissent,* 68–69; see also "Thomas Jefferson to Samuel Latham Mitchill," June 13, 1800, in McClure and Looney, *Papers of Thomas Jefferson.*

33. For Genet, see Elkins and McKitrick, *Age of Federalism,* especially 341–54. For Michaux, see Hayes, *Road to Monticello,* 413–15; and especially Elkins and McKitrick, *Age of Federalism,* 349–51.

34. Dumas Malone, *Jefferson and the Ordeal of Liberty* (New York: St. Martin's Press, 1962), 104–9.

35. For tattooing, see Hayes, *Road to Monticello,* 413–15. For the subscription, see Michaux Subscription, American Philosophical Society.

36. Hayes, *Road to Monticello,* 413–15; Malone, *Ordeal of Liberty,* 107–8; and Boles, *Architect of American Liberty,* 251–52.

37. "Circular Letter: The Society Having Appointed a Committee to Collect Information Respecting the Past and Present State of This Country, the Committee during the Last Year Addressed the following Letter to Such Persons as Were Likely, in

Their Opinion to Advance the Object of the Society," *Transactions of the American Philosophical Society* 4 (1799): xxxvii–xxxix. For internal details of the committee, of which there are only a few references, see "APS Minutes, 1793–1798"; see also "APS Minutes, 1799–1804," *APS Library*, http://diglib.amphilsoc.org/islandora/object/american-philosophical-society-minutes-1799-1804.

38. "Circular Letter," xxxvii–xxxix. For examples of reports arriving as part of this APS initiative, see "Benjamin Hawkins to Thomas Jefferson," July 12, 1800, in McClure and Looney, *Papers of Thomas Jefferson.*

39. For more on Jefferson's interest in Native languages and cultures and how it fit into his political worldview, see Kerber, *Federalists in Dissent,* 71; see also Thomson, *Jefferson's Shadow,* 101–107.

40. See Thomson, *Jefferson's Shadow,* especially 101–7.

41. Thomas Jefferson, *Notes on the State of Virginia* (Boston: Sprague, 1802), 87–88; Thomson, *Jefferson's Shadow,* 101–107.

42. "Thomas Jefferson to Benjamin Smith Barton," September 21, 1809, in McClure and Looney, *Papers of Thomas Jefferson.*

43. "APS Minutes, 1793–1798"; "Memoir on the Megalonyx [10 February 1797]," *Founders Online,* National Archives, last modified April 12, 2018, http://founders. archives.gov/documents/Jefferson/01-29-02-0232 [Original source: Barbara B. Oberg, ed., *The Papers of Thomas Jefferson,* vol. 29, *1 March 1796–31 December 1797* (Princeton, NJ: Princeton University Press, 2002), 291–304]; and Thomson, *Jefferson's Shadow,* 87–97.

44. For weights and measures, see "Thomas Jefferson to David Rittenhouse," June 12, 1790, June 15, 1790, June 20, 1790, June 26, 1790, and June 30, 1790; and "Rittenhouse to Jefferson," June 21, 1790 and July 2, 1790, all in McClure and Looney, *Papers of Thomas Jefferson.* For an example of Jefferson recommending correspondents, see "Thomas Jefferson to Dugald Stewart," June 2, 1797; and "Thomas Jefferson to Caspar Wistar," December 16,

1800 (nominating Dunbar), both in McClure and Looney, *Papers of Thomas Jefferson*. For viniculture, see Jefferson's correspondence with Peter Legaux and the Journal of the Vine Company of Philadelphia, American Philosophical Society Library.

45. Data is taken from the APS Minutes, available at http://diglib.amphilsoc.org/islandora/graphics/minutes-american-philosophical-society. Full datasets will be released upon completion of the project. See "Thomas Jefferson to John Stuart," November 19, 1796, in McClure and Looney, *Papers of Thomas Jefferson*.

46. Kerber, *Federalists in Dissent*, 67–94, quotes on 79 and 77.

47. This data is based on analysis on the subject of articles in the *Transactions of the American Philosophical Society* (accessed via JSTOR) between 1786 and 1839.

48. "Thomas Jefferson to Thomas James," January 21, 1803; and "Thomas Jefferson to John Vaughan," January 14, 1802, both in McClure and Looney, *Papers of Thomas Jefferson*.

49. Data here and below comes from an analysis of the *Jefferson Papers*. General correspondence to APS Members on unrelated business was excluded. For instance, George Washington was a frequent recipient of letters from Jefferson in the 1790s, but only those that dealt with APS business were included. A full dataset will be released upon completion of the project.

50. For correspondence on vaccination, see "John Vaughan to Thomas Jefferson," November 20, 1801, December 21, 1801, April 28, 1803, and December 20, 1803; and "John Redman Coxe to Thomas Jefferson," January 1, 1803, all in McClure and Looney, *Papers of Thomas Jefferson*. For ciphering, see "Method of Using Robert Patterson's Cipher" in McClure and Looney, *Papers of Thomas Jefferson*.

51. For the best biography of Ellicott, see William J. Morton, *Andrew Ellicott: The Stargazer Who Defined America* (Atlanta: Georgia History Press, 2015).

52. For more on Smith, see "Thomas Jefferson to Samuel Harrison Smith," June 24, 1800, in McClure and Looney, *Papers of Thomas Jefferson*; and Samuel Harrison Smith, *Remarks on Education: Illustrating the Close Connection Between Virtue and Wisdom: To Which Is Annexed, a System of Liberal Education*, accessed May 6, 2018, https://link.springer.com/chapter/10.1057/9781137271020_12.

53. "Thomas Jefferson to Andrew Ellicott," December 18, 1800, in McClure and Looney, *Papers of Thomas Jefferson*.

54. For an offer of assistance, see "Thomas Jefferson," July 29, 1801, and "Charles Willson Peale to Thomas Jefferson," October 11, 1801, in McClure and Looney, *Papers of Thomas Jefferson*.

55. "To Thomas Jefferson from Jonathan Williams, 3 February 1806," *Founders Online*, National Archives, accessed April 11, 2019, https://founders.archives.gov/documents/ Jefferson/99-01-02-3173; "To Thomas Jefferson from Jonathan Williams, 12 December 1802," *Founders Online*, National Archives, last modified June 13, 2018, http://founders.archives.gov/documents/Jefferson/01-39-02-0134 [Original source: Barbara B. Oberg, ed., *The Papers of Thomas Jefferson*, vol. 39, *13 November 1802–3 March 1803* (Princeton, NJ: Princeton University Press, 2012), 145–46]; "From Thomas Jefferson to Jonathan Williams, 25 December 1802," *Founders Online*, National Archives, last modified June 13, 2018, http://founders.archives.gov/documents/Jefferson/01-39-02-0199 [Original source: Oberg, *Papers of Thomas Jefferson*, vol. 39, 220–21].

56. "From Thomas Jefferson to Jonathan Williams, 14 July 1805," *Founders Online*, National Archives, last modified June 13, 2018, http://founders.archives.gov/documents/Jefferson/99-01-02-2096.

57. "Thomas Jefferson to John Vaughan," June 7, 1817; and "John Vaughan to Thomas Jefferson," July 30, 1817, both in McClure and Looney, *Papers of Thomas Jefferson*. A dataset on

Jefferson's exchanges will be published upon completion of the project.

58. Special thanks to Andrew Fagal for sharing this exchange with us. The letter is yet to be published, but the Society's response can be found in "APS Minutes, 1805–1814," *APS Library*, accessed May 6, 2018, http://diglib.amphilsoc.org/islandora/object/ameri can-philosophical-society-minutes-1805-1814.

59. The 40 percent comes from an analysis of the APS Member Directory from 1743 to 1826.

60. "Thomas Jefferson to Thomas Hewson," January 21, 1810; and "Thomas Hewson to Thomas Jefferson," January 8, 1810, both in McClure and Looney, *Papers of Thomas Jefferson*.

61. The frequency of correspondence is based on the *Founders Online* ranking of frequent correspondents.

62. "Thomas Jefferson to Peter S. Du Ponceau, 22 January 1816," *Founders Online*, National Archives, last modified June 13, 2018, http://founders.archives.gov/documents/Jefferson/03-09-02-0246 [Original source: J. Jefferson Looney, ed., *The Papers of Thomas Jefferson*, Retirement Series, vol. 9, *September 1815 to April 1816* (Princeton, NJ: Princeton University Press, 2012), 383–84.]

63. "Thomas Jefferson to John Vaughan," June 7, 1817; and "John Vaughan to Thomas Jefferson," July 30, 1817, both in McClure and Looney, *Papers of Thomas Jefferson*.

64. "Jonathan Williams to Thomas Jefferson, 21 October 1814," *Founders Online*, Jefferson/03-08-02-0033 [Original source: J. Jefferson Looney, ed., *The Papers of Thomas Jefferson*, Retirement Series, vol. 8, *1 October 1814 to 31 August 1815* (Princeton, NJ: Princeton University Press, 2011), 40–41.]

65. "From University of Virginia Board of Visitors to Literary Fund Board, 29 November 1821," *Founders Online*, National Archives, last modified June 13, 2018, http://founders. archives .gov/documents/Jefferson/98-01-02-2460.

66. "From Thomas Jefferson to Robert Patterson, 8 July 1823," *Founders Online,* National Archives, last modified June 13, 2018, http://founders.archives.gov/documents/Jefferson/98-01-02-3620.

67. "Memorandum Books, 1824," *Founders Online*, National Archives, last modified June 13, 2018, http://founders.archives.gov/documents/Jefferson/02-02-02-0034 [Original source: James A Bear Jr. and Lucia C. Stanton, eds., *The Papers of Thomas Jefferson,* Second Series, *Jefferson's Memorandum Books,* vol. 2 (Princeton, NJ: Princeton University Press, 1997), 1401–408]; "From Thomas Jefferson to John Vaughan, 24 March 1826," *Founders Online*, National Archives, last modified June 13, 2018, http://founders.archives.gov/documents/Jefferson/98-01-02-5982.

Index

Index

Index

Index

Index